I0489913

Starship, Our Ticket to the Universe

By Alex Samuelson

First Edition March 2023
ISBN: 9798388424273

Photos by Cosmic Perspective

This book is dedicated to my wife and son.
May their future be bright and they awake each day with
something to look forward to.

Starship, Our Ticket to the Universe

"When something is important enough, you do it even if the odds are not in your favor."
Elon Musk

Chapter 1: Introduction to Starship and the New Space Age

The emergence of Starship technology

SpaceX's Starship spacecraft represents a significant milestone in the history of space exploration. Designed to transport people and cargo to the Moon, Mars, and beyond, Starship is poised to revolutionize the way we explore and settle the solar system. But how did Starship technology emerge, and what makes it so unique?

The early days of SpaceX were focused on developing reusable rocket technology. In 2015, SpaceX successfully landed its Falcon 9 rocket, demonstrating for the first time that a rocket could be reused after a successful launch. This breakthrough paved the way for the development of even more advanced rapidly reusable rocket technology, including the Starship spacecraft. According to SpaceX's founder and Chief Engineer Elon Musk, *"With respect to space, I think there's really just one problem, which is a fully and rapidly reusable orbital rocket. This is the holy grail."* This goal is fulfilled by the Starship system.

The Starship spacecraft is a fully reusable, two-stage vehicle that can carry up to 100 people or 100 metric tons of cargo to destinations throughout the solar system. The spacecraft consists of two parts: the Starship vehicle, which serves as the passenger and cargo module, and the Super Heavy rocket, which provides the initial boost needed to launch the Starship into orbit.

What makes Starship technology so unique is its use of advanced materials and engineering solutions to reduce weight and increase performance. The spacecraft is constructed primarily of stainless steel, which is both strong and lightweight. The design of the spacecraft also includes several innovative features, such as a lack of landing legs (it is intended to return to the launch pad) and a heat shield made of silica fiber tiles.

One of the most significant advantages of Starship technology is its reusability. Unlike previous spacecraft, which were used only once before being discarded, Starship is intended to be rapidly flown multiple times, significantly reducing the cost of spaceflight. This is made possible by the spacecraft's advanced engineering, which allows it to withstand the stresses of spaceflight and re-entry into Earth's atmosphere.

Another advantage of Starship technology is its versatility. The spacecraft is designed to be modular, meaning that it can be configured to carry passengers, cargo, or scientific instruments. This flexibility makes Starship an ideal platform for a wide range of missions, from crewed missions to Mars to robotic missions to explore the outer planets.

The emergence of Starship technology represents a significant milestone in the history of space exploration. With its advanced materials, innovative design, and unprecedented reusability, Starship is poised to revolutionize the way we explore and settle the solar system. While the development of this technology has not been without its challenges, the potential benefits of Starship technology are immense, from scientific discovery to economic growth to the expansion of human civilization beyond Earth. As we look to the future, it is clear that Starship technology will play a critical role in the ongoing exploration and settlement of the solar system.

The new era of space exploration and colonization

With the emergence of new space technologies like Starship, we are on the cusp of a new era of space exploration and colonization. While previous missions to space have been focused on scientific research and exploration, the new era of space exploration will focus on human settlement and the expansion of civilization beyond Earth.

One of the primary drivers of this new era of space exploration is the growing realization that Earth's resources are finite. As our population grows and our technological capabilities expand, we are putting increasing pressure on our planet's resources. The idea of space colonization has long been seen as a way to alleviate this pressure by providing access to new sources of energy, materials, and living space.

Another factor driving the new era of space exploration is the growing recognition of the importance of space for national security and economic competitiveness. Countries around the world are investing heavily in space technology and exploration, recognizing that space provides critical advantages in areas such as communication, intelligence gathering, and scientific research.

Perhaps the most significant driver of the new era of space exploration, however, is the desire to explore and discover. Humans have an innate curiosity and desire to explore the unknown, and space is the ultimate frontier. The potential for scientific discovery and exploration is vast, with the potential to uncover new knowledge about our universe and our place in it.

But what will this new era of space exploration and colonization look like? In many ways, it will be similar to previous eras of exploration, with new frontiers to explore and new challenges to overcome. But there will also be significant differences, such as the need to develop new technologies for long-duration spaceflight, the challenge of creating self-sustaining habitats in space, and the ethical and social considerations of creating a new civilization beyond Earth.

Despite these challenges, the potential benefits of space exploration and colonization are immense. In addition to providing access to new resources and scientific discoveries, space colonization could also serve as a means of ensuring the long-term survival of humanity. By creating a self-sustaining civilization beyond Earth, we can ensure that our species continues to thrive even in the face of global catastrophes.

The new era of space exploration and colonization represents a significant turning point in human history. With new technologies like Starship, we are poised to explore new frontiers and create a new civilization beyond Earth. While the challenges are significant, the potential rewards are even greater, providing a pathway to a brighter and more prosperous future for all of humanity.

The impact of Starship on space travel and human civilization

Space travel has always captured the imagination of people around the world. For decades, we have dreamed of exploring the universe, colonizing other planets, and discovering new worlds. The development of Starship by SpaceX represents a major milestone in the history of space travel, with the potential to revolutionize the way we explore and colonize space.

One of the most significant impacts of Starship is its potential to reduce the cost of space travel. Traditional rockets, like the Space Shuttle, are incredibly expensive to build and maintain, making space travel prohibitively expensive for most people and organizations. Starship, on the other hand, is designed to be reusable, with the ability to land on Earth or other planets and then take off again, dramatically reducing the cost of space travel.

This reduction in cost could lead to a number of new opportunities for space exploration and colonization. For example, it could make it possible for private companies and individuals to launch their own missions to space, opening up new frontiers for exploration and discovery. It could also make it possible to establish permanent human settlements on other planets, something that has long been seen as a key goal of space exploration.

Another significant impact of Starship is its potential to create a new era of international cooperation in space. Historically, space travel has been dominated by a few major players, primarily the United States and Russia. However, with the development of Starship, it is possible that other countries will begin to develop their own space programs, leading to increased international cooperation and collaboration in space exploration.

The development of Starship could also have significant impacts on the way we live our lives here on Earth. For example, space-based technologies could be used to solve some of our most pressing environmental and societal challenges. From climate change to energy production to transportation, space-based technologies have the potential to revolutionize the way we live our lives and create a more sustainable future for all of humanity.

Finally, the development of Starship and other space technologies could inspire a new generation of scientists, engineers, and explorers. By showing that it is possible to explore and colonize space, we can inspire young people to pursue careers in science and technology, driving innovation and progress in a wide range of fields.

The development of Starship represents a major milestone in the history of space travel and exploration. Its potential to reduce the cost of space travel, create new opportunities for exploration and colonization, and inspire a new generation of scientists and explorers could have significant impacts on the future of human civilization. Whether we are exploring new worlds, solving our most pressing challenges, or inspiring the next generation of innovators, the impact of Starship on space travel and human civilization is sure to be profound.

"Rockets are cool. There's no getting around that."

Elon Musk

Chapter 2: The History of Starship

The origins of Starship technology

The development of Starship by SpaceX has been a major milestone in the history of space exploration. However, the technology behind Starship did not emerge overnight. In fact, it has its roots in a long history of research and development, much of which was focused on the goal of making space travel more accessible and affordable.

One of the earliest precursors to Starship technology was the Saturn V rocket, developed by NASA in the 1960s. The Saturn V was designed to carry the Apollo spacecraft to the moon, and was one of the most powerful rockets ever built. It consisted of three stages, each of which was designed to be discarded after use. This approach was later adapted by SpaceX, who developed the Falcon 9 rocket with a similar multi-stage design. Only the first stage was made reusable on the Falcon 9.

In addition to the Saturn V, another important precursor to Starship was the Space Shuttle. The Space Shuttle was designed to be a reusable spacecraft, with the ability to launch into orbit and then return to Earth for reuse. Although the Space Shuttle was not as successful as initially hoped, it did pave the way for the development of reusable spacecraft technology, which has since been adapted by SpaceX for use in Starship.

Another important influence on the development of Starship technology was the X-33 program, developed by NASA in the 1990s. The X-33 was a prototype spacecraft designed to test a number of new technologies, including composite materials and a new type of rocket engine. Although the X-33 program was ultimately cancelled, it did help to advance the state of the art in spacecraft design and propulsion, providing important lessons for SpaceX and other companies working on the development of new spacecraft technologies.

Despite these early precursors, the development of Starship by SpaceX represents a major leap forward in the state of the art in space travel technology. With its revolutionary design and reusable capabilities, Starship has the potential to dramatically reduce the cost of space travel and open up new frontiers for exploration and colonization. Moreover, the development of Starship has inspired a new wave of interest in space travel and exploration, paving the way for new discoveries and innovations in the years to come.

The development of Starship technology is the result of a long and rich history of research and development in the field of space travel. Although its origins can be traced back to the Saturn V, the Space Shuttle, and the X-33 programs, the development of Starship by SpaceX represents a major leap forward in the state of the art in spacecraft technology. Its innovative design and reusable capabilities have the potential to revolutionize the way we explore and colonize space, inspiring a new era of space travel and exploration for generations to come.

The development of Starship technology over time

The development of Starship technology by SpaceX has been a complex and challenging process, requiring the integration of numerous cutting-edge technologies and the coordination of a large and highly skilled team of engineers, scientists, and technicians. From its initial conception to its current status as one of the most ambitious and innovative spacecraft designs ever created, the story of Starship is one of perseverance, creativity, and technical excellence.

The origins of Starship can be traced back to the early days of SpaceX, when company founder Elon Musk first began to explore the possibilities of space travel and colonization. In 2011, Musk announced his vision for a new spacecraft design that would be capable of carrying humans and cargo to Mars and other destinations in the solar system. This vision would eventually take shape as the Starship program.

The early years of Starship development were characterized by a series of technical challenges and setbacks. The first iteration of the spacecraft, known as the ITS (Interplanetary Transport System), was initially envisioned as a massive, 122-meter-tall rocket with the capability to carry up to 450 metric tons of cargo to Mars. However, after several years of design and testing, SpaceX eventually realized that this approach was not feasible, and instead shifted its focus to a new design that would be smaller, more affordable, and more versatile.

This new design, initially called the BFR (Big Falcon Rocket) but which would eventually be known as the Starship, was unveiled by SpaceX in 2019. It consists of a 70 meter tall booster called Super Heavy, and a upper stage called Starship. Fully stacked, it stands 120 meters tall. Starship is a 50-meter-tall spacecraft with a diameter of 9 meters, capable of carrying up to 100 metric tons of cargo or 100 passengers to Mars and other destinations in the solar system. Initial planning had Starship made of carbon fiber, but problems with this material caused a switch to stainless steel. One of the key features of the Starship design is its use of stainless steel, a material that is both strong and lightweight, and which allows the spacecraft to withstand the extreme conditions of space travel.

The development of Starship has also been marked by a series of successful test flights and demonstrations. In 2020 and 2021, SpaceX conducted a series of tests of the Starship prototype, including several high-altitude flights that demonstrated the spacecraft's ability to take off, maneuver, and land safely. Although there have been some setbacks and technical challenges along the way, SpaceX has remained committed to the development of Starship, with plans to continue testing and refining the spacecraft in the years to come.

Looking to the future, the development of Starship technology is likely to continue to evolve and change in response to new discoveries and technological advancements. SpaceX has already announced plans to use the Starship for a variety of missions, including lunar and Mars exploration, space tourism, and satellite deployment. Moreover, the development of Starship has inspired a new wave of interest and investment in space exploration and colonization, paving the way for new discoveries and innovations in the years to come.

The development of Starship technology by SpaceX has been a long and challenging process, requiring the integration of numerous cutting-edge technologies and the coordination of a large and highly skilled team of engineers, scientists, and technicians. Despite the setbacks and challenges, SpaceX has remained committed to the development of Starship, with plans to continue testing and refining the spacecraft in the years to come. As the technology continues to evolve and improve, it has the potential to revolutionize the way we explore and colonize space, paving the way for new discoveries and innovations in the decades to come.

The role of Starship in human space exploration

The development of Starship technology has the potential to revolutionize human space exploration, paving the way for new discoveries, advancements, and possibilities. Starship is a versatile spacecraft that has been designed to support a wide range of missions, including lunar and Mars exploration, satellite deployment, space tourism, and more. Its capabilities and potential uses have garnered significant interest and investment from governments, private companies, and individuals around the world.

One of the key roles of Starship in human space exploration is to enable more efficient and cost-effective access to space. The spacecraft's innovative design, which includes reusable rockets and spacecraft, could significantly reduce the cost of launching payloads and people into space. This could open up new opportunities for space exploration and research, as well as for commercial space activities such as satellite deployment and space tourism.

Another important role of Starship is to facilitate the colonization of space. The spacecraft's capabilities make it a promising tool for establishing human settlements on the Moon, Mars, and other destinations in the solar system. The ability to transport large payloads and a significant number of people could help to support long-term missions and to establish self-sustaining colonies in space. The potential benefits of space colonization include expanding the resources available to humanity, providing new opportunities for scientific research, and ensuring the survival of the human species in the event of a catastrophic event on Earth.

Starship technology also has the potential to facilitate international cooperation and collaboration in space exploration. The development of the spacecraft has already attracted significant interest and investment from governments and private companies around the world, indicating a growing global commitment to space exploration and research. The ability to share resources, knowledge, and expertise could help to accelerate the pace of space exploration and to achieve new discoveries and advancements.

Moreover, the development of Starship technology has the potential to inspire and engage the public in space exploration and science. The prospect of human missions to Mars and other destinations in the solar system has captured the public imagination and generated significant interest and excitement. The development of Starship technology, and the missions and discoveries it enables, could help to inspire a new generation of scientists, engineers, and space enthusiasts.

The role of Starship in human space exploration is multifaceted and promising. The spacecraft's innovative design and capabilities could significantly reduce the cost and increase the efficiency of space exploration and research. It could also facilitate the colonization of space, provide new opportunities for international cooperation and collaboration, and inspire and engage the public in space exploration and science. As the development of Starship technology continues, it has the potential to unlock new discoveries, advancements, and possibilities in the realm of human space exploration.

"If we drive down the cost of transportation in space, we can do great things."

Elon Musk

Chapter 3: The Mechanics of Starship

The design and construction of Starship

The Starship spacecraft is the result of years of research, development, and testing by SpaceX, a private space company founded by entrepreneur Elon Musk. The innovative design of the spacecraft is intended to provide a reusable, cost-effective, and versatile platform for a wide range of space missions, including exploration, research, and commercial activities.

The design of Starship is centered around a fully reusable, two-stage rocket system. The first stage is known as the Super Heavy, a massive rocket booster that is designed to launch the Starship spacecraft into space. The Super Heavy booster is powered by 33 Raptor engines, which are fueled by methane and oxygen. This configuration provides a high level of thrust and allows the Starship to lift off with a payload capacity of up to 150 metric tons in a reusable configuration.

The second stage of the Starship spacecraft is known as the Starship itself, which is designed to carry crew and cargo to and from space. The Starship spacecraft features a stainless steel construction, which provides strength, durability, and resistance to high temperatures during re-entry into Earth's atmosphere. The spacecraft is powered by six Raptor engines, which are also fueled by methane and oxygen. The Starship spacecraft has a payload capacity of up to 150 metric tons reusable (250 tons expendable) and can carry up to 100 people into space. Up to 1000 people can be accommodated on a shorter hop between locations on Earth.

One of the key features of the Starship design is its reusability. The Super Heavy booster and Starship spacecraft are both designed to be fully reusable, allowing for multiple launches and landings with minimal maintenance and refurbishment. The reusable design is intended to reduce the cost of space launches and make space exploration and research more accessible and sustainable.

The construction of the Starship spacecraft involves a combination of traditional manufacturing techniques and innovative technologies. The stainless steel construction of the spacecraft is fabricated using automated welding techniques, which allow for precise and efficient assembly of the spacecraft components. The spacecraft is also outfitted with a range of advanced technologies, including avionics systems, communication systems, and life support systems.

The construction and testing of the Starship spacecraft have involved a significant amount of research and development, as well as extensive testing and prototyping. SpaceX has conducted a series of test flights and ground tests to validate the performance and safety of the spacecraft. These tests have included low-altitude hops, high-altitude flights, and prototype failures, which have helped to refine the design and identify areas for improvement.

The design and construction of Starship is a complex and innovative process that has resulted in a versatile, reusable, and cost-effective spacecraft platform. The design of the spacecraft, with its stainless steel construction, automated welding techniques, and advanced technologies, provides a robust and reliable platform for space exploration and research. The reusability of the spacecraft is a key feature that is intended to make space exploration more sustainable and accessible, and the extensive testing and development process has helped to ensure the safety and reliability of the Starship spacecraft.

The propulsion and navigation systems of Starship

The propulsion and navigation systems of Starship are critical components of the spacecraft, providing the necessary power and control for safe and efficient space travel. The Starship spacecraft features advanced propulsion systems that enable the spacecraft to travel long distances and maneuver in space, as well as advanced navigation systems that provide accurate positioning and guidance.

The propulsion system of Starship is based on the Raptor engines, which are fueled by methane and liquid oxygen. The spacecraft is equipped with six Raptor engines (3 each for Sea Level and Vacuum), which provide a total thrust of approximately 1,500 tons. The engines are arranged in a symmetric pattern around the base of the spacecraft, providing balanced and controlled thrust during launch and landing.

The Raptor engines are designed to be highly efficient and reliable, with advanced technology that enables the engines to operate at high temperatures and pressures. The engines feature a regenerative cooling system, which uses the fuel to cool the engine parts and reduce the risk of overheating. This design allows the Raptor engines to achieve high levels of performance while maintaining a high degree of safety and reliability.

The propulsion system of Starship also includes a set of control thrusters, which are used for maneuvering and stabilization during flight. The control thrusters are located at the top and bottom of the spacecraft, providing a high degree of control over the spacecraft's orientation and trajectory.

The navigation system of Starship is based on a set of advanced sensors and computers, which provide accurate positioning and guidance for the spacecraft. The spacecraft is equipped with a set of inertial navigation sensors, which use accelerometers and gyroscopes to measure the spacecraft's position, velocity, and orientation.

The spacecraft is also equipped with a set of star trackers, which use the stars to determine the spacecraft's position and orientation. The star trackers are highly accurate and provide a reliable backup to the inertial navigation system.

The navigation system of Starship is controlled by a set of advanced computers, which process the sensor data and provide real-time guidance and control for the spacecraft. The computers are equipped with advanced algorithms and software, which enable the spacecraft to navigate in space with a high degree of accuracy and precision.

The propulsion and navigation systems of Starship are critical components that enable the spacecraft to travel long distances and maneuver in space. The Raptor engines provide the necessary thrust and power for safe and efficient space travel, while the control thrusters provide a high degree of control and maneuverability. The navigation system of Starship is based on advanced sensors and computers, which provide accurate positioning and guidance for the spacecraft. The propulsion and navigation systems of Starship are key features that enable the spacecraft to achieve its mission objectives and explore the depths of space.

The materials and technologies used in building Starship

The construction of Starship involves the use of advanced materials and technologies, which enable the spacecraft to withstand the harsh conditions of space travel and exploration. The materials and technologies used in building Starship are carefully selected and tested to ensure the highest levels of safety and reliability.

One of the key materials used in building Starship is stainless steel, which is a highly durable and heat-resistant material. Stainless steel is used for the spacecraft's primary structure, including the hull and the tanks that hold the fuel and oxidizer. The use of stainless steel in the construction of Starship provides several advantages, including high strength, low weight, and resistance to corrosion. The material is also durable, and is likely to give a Starship a reasonable life span.

Another key material used in building Starship is silica fiber, which is a durable and lightweight material. Silica fiber is used in the spacecraft's heat shield, which protects the spacecraft from the high temperatures of reentry into Earth's atmosphere. The use of silica fiber in the heat shield enables the spacecraft to withstand the intense heat of reentry while maintaining a low weight.

In addition to stainless steel and silica fiber, Starship also incorporates advanced technologies and manufacturing processes. One of these technologies is friction stir welding, which is a process that enables the joining of metals without the need for traditional welding techniques. Friction stir welding is used in the construction of Starship to join the stainless steel panels that make up the spacecraft's hull.

Another key technology used in building Starship is 3D printing, which enables the rapid and efficient manufacturing of complex parts and components. 3D printing is used in the construction of the Raptor engines, which are made up of hundreds of intricate parts. The use of 3D printing in the manufacturing of the Raptor engines enables the engines to be produced with a high degree of precision and reliability.

The use of advanced materials and technologies in the construction of Starship enables the spacecraft to achieve its mission objectives and explore the depths of space with a high degree of safety and reliability. The materials and technologies used in building Starship are carefully selected and tested to ensure that the spacecraft can withstand the extreme conditions of space travel and exploration.

The complete Starship itself is not the only component in the Starship system, however. Along with all the construction and support devices, the launching structure itself is integral to the use of the rocket. The launch structure, consisting of the Orbital Launch Mount and the chopsticks-equipped tower called "Mechazilla", is also intended to be used for landings. The iconic tower of the launch structure includes devices called "chopsticks" which jut out to run up and down the tower on tracks, and both seat the booster and Starship upper stage for launching and catch them on their return flight. This is an integral part of the entire Starship system, as the Starship upper stage and Super Heavy booster do not include any landing legs and must be caught by the chopsticks on their return to the launch site.

The construction of Starship involves the use of advanced materials and technologies, including stainless steel, silica fiber, friction stir welding, and 3D printing. These materials and technologies enable the spacecraft to achieve its mission objectives and explore the depths of space with a high degree of safety and reliability. The use of advanced materials and technologies in building Starship represents a major milestone in the development of space exploration and the exploration of the universe beyond our planet.

"If it were to take longer to convince NASA and the authorities that we can do it versus just doing it, then we might just do it. It may literally be easier to just land Starship on the moon than try to convince NASA that we can."

Elon Musk

Chapter 4: The Advantages of Starship

The speed and efficiency of Starship travel

Starship is designed to travel at incredible speeds and with a high degree of efficiency, making it a crucial tool for future space exploration missions. With its advanced propulsion system and aerodynamic design, Starship is capable of achieving speeds that were once considered impossible.

One of the key features of Starship's propulsion system is the use of Raptor engines. These engines are designed to provide a high degree of thrust, enabling the spacecraft to accelerate quickly and reach high speeds. The Raptor engines use a combination of methane and liquid oxygen as fuel and oxidizer, which provides a high specific impulse, or efficiency, compared to other rocket engines.

In addition to its advanced propulsion system, Starship also incorporates an aerodynamic design that reduces drag and maximizes efficiency. The spacecraft's shape is optimized to reduce the amount of fuel needed to achieve and maintain its high speeds. SpaceX has pioneered a new type of wing to use atmospheric resistance in reentry. The use of a heat shield made from silica fiber also helps to reduce unwanted drag and increase efficiency, as it enables the spacecraft to withstand the high temperatures of reentry without suffering damage.

The speed and efficiency of Starship travel enable the spacecraft to complete missions in a shorter amount of time, while using less fuel and resources. This is particularly important for long-duration missions, such as missions to Mars and other planets, where minimizing travel time and maximizing efficiency is crucial to the success of the mission.

Another important factor in the speed and efficiency of Starship travel is the spacecraft's ability to refuel in orbit. Starship is designed to carry a large amount of fuel and oxidizer, which enables it to travel long distances. However, by refueling in orbit, the spacecraft can extend its range and travel even further without the need for additional fuel.

The speed and efficiency of Starship travel represent a major milestone in the development of space exploration and the ability of humans to explore the universe beyond our planet. With its advanced propulsion system, aerodynamic design, and the ability to refuel in orbit, Starship is poised to become a crucial tool in future space exploration missions, enabling humans to travel further and faster than ever before.

Starship's speed and efficiency represent a major leap forward in space exploration technology. With its advanced propulsion system, aerodynamic design, and ability to refuel in orbit, Starship is capable of achieving high speeds and traveling long distances with a high degree of efficiency. This enables the spacecraft to complete missions in a shorter amount of time, while using less fuel and resources, making it a crucial tool in future space exploration missions.

The ability of Starship to transport large numbers of people and cargo

One of the key advantages of Starship is its ability to transport large numbers of people and cargo into space. With its massive size and advanced design, the spacecraft is capable of carrying significant amounts of material and personnel to destinations beyond our planet.

Starship's cargo capacity is particularly impressive. The spacecraft can carry up to 100 metric tons of cargo, including supplies, equipment, and even entire habitats for human settlements on other planets. This makes Starship an ideal tool for resupply missions to the International Space Station (ISS) and other space habitats, as well as for carrying equipment and supplies for future space exploration missions.

In addition to its cargo capacity, Starship is also designed to transport large numbers of people. The spacecraft's interior can be configured to accommodate up to 100 passengers, making it an ideal tool for transporting astronauts and researchers to destinations beyond our planet. For destinations on our planet (and short hops), up to 1000 passengers can fit in a Starship's hull.

The ability of Starship to transport large numbers of people and cargo has significant implications for the future of space exploration and colonization. With the ability to transport significant amounts of material and personnel, Starship could facilitate the establishment of human settlements on other planets and enable humans to explore the universe beyond our planet.

In addition to its ability to transport large numbers of people and cargo, Starship is also designed to be reusable. This means that the spacecraft can be used for multiple missions, reducing the cost of space exploration and making it more accessible to researchers, entrepreneurs, and governments around the world.

The ability of Starship to transport large numbers of people and cargo represents a major milestone in the development of space exploration technology. With its massive size and advanced design, the spacecraft is capable of carrying significant amounts of material and personnel to destinations beyond our planet, making it a crucial tool in future space exploration and colonization missions.

Starship's ability to transport large numbers of people and cargo has significant implications for the future of space exploration and colonization. With the ability to transport significant amounts of material and personnel, the spacecraft could facilitate the establishment of human settlements on other planets and enable humans to explore the universe beyond our planet. The ability of Starship to transport large numbers of people and cargo, combined with its reusability, represents a major milestone in the development of space exploration technology and makes it a crucial tool in future space exploration missions.

The potential for Starship to facilitate human colonization of other planets

One of the most exciting possibilities presented by Starship technology is the potential for human colonization of other planets. With the spacecraft's ability to transport large numbers of people and cargo, and its reusability, Starship could play a crucial role in establishing permanent human settlements on other planets.

The first step in human colonization of other planets is identifying viable destinations for settlement. Mars is one of the most promising options, with its similar size, gravity, and atmosphere to Earth, as well as evidence of water on its surface. Other potential destinations for settlement include the Moon, Venus (at least the upper atmosphere), and some of the moons of Jupiter and Saturn.

Once a destination has been selected, Starship could be used to transport personnel and equipment to the planet or moon. The spacecraft's cargo capacity and interior space could be used to transport everything needed to establish a human settlement, including supplies, equipment, habitats, and even greenhouses for growing crops.

One of the major challenges in human colonization of other planets is the difficulty of resupply missions. With Starship's ability to carry significant amounts of material and personnel, the spacecraft could help alleviate some of the challenges associated with resupply missions. In addition, the spacecraft's reusability could significantly reduce the cost of establishing and maintaining a human settlement on another planet.

The potential for Starship to facilitate human colonization of other planets is an exciting possibility that could revolutionize our understanding of the universe and our place within it. With its ability to transport large numbers of people and cargo, and its reusability, Starship could be a crucial tool in establishing permanent human settlements on other planets.

However, there are still significant challenges to be overcome in human colonization of other planets. These include developing technologies to sustain life on other planets, addressing radiation exposure and other health risks, and establishing a self-sustaining ecosystem. Despite these challenges, the potential benefits of human colonization of other planets, including the expansion of human knowledge and the potential for new resources and discoveries, make it a worthwhile pursuit.

The potential for Starship to facilitate human colonization of other planets is an exciting possibility that could revolutionize our understanding of the universe and our place within it. With its ability to transport large numbers of people and cargo, and its reusability, Starship could be a crucial tool in establishing permanent human settlements on other planets. While significant challenges remain, the potential benefits of human colonization of other planets make it a worthwhile pursuit for the future of humanity.

"I would like to die on Mars. Just not on impact."

Elon Musk

Chapter 5: The Challenges of Starship

The cost and complexity of building and operating Starship

While Starship presents a promising future for space exploration and colonization, building and operating the spacecraft is an expensive and complex endeavor. From the design and construction to the launch and operation, Starship requires significant resources and expertise.

The cost of building and operating Starship is high due to the complex design and technology involved. The spacecraft's stainless steel construction, coupled with advanced propulsion and navigation systems, require a significant investment in research and development. Additionally, the manufacturing process for Starship is complex and time-consuming, with each spacecraft requiring hundreds of thousands of parts and specialized equipment.

The operational costs of Starship are also significant. The spacecraft requires a specialized launch infrastructure, including launch pads and fueling systems, which are expensive to construct and maintain. Additionally, the cost of fueling Starship must be considered, with the spacecraft requiring large amounts of propellant for each launch.

Despite the high costs, SpaceX has continued to invest in the development and operation of Starship, recognizing the potential benefits of the technology. The reusability of the spacecraft is one factor that could help reduce the overall cost of building and operating Starship, as each spacecraft can be used for multiple launches.

Another potential factor that could help reduce the cost of Starship is the development of new technologies and manufacturing techniques. For example, SpaceX has invested in a new type of manufacturing technology known as "friction stir welding," which could significantly reduce the time and cost of building Starship. Over time, the production of Starships will evolve and change.

Overall, while the cost and complexity of building and operating Starship are significant challenges, the potential benefits of the technology, including its ability to revolutionize space exploration and colonization, make it a worthwhile investment for SpaceX and the future of humanity's exploration of space.

The cost and complexity of building and operating Starship are significant challenges for SpaceX and the development of the technology. The spacecraft's complex design and advanced technology require a significant investment in research and development, as well as specialized manufacturing and launch infrastructure. However, the potential benefits of Starship, including its ability to revolutionize space exploration and colonization, make it a worthwhile investment for the future of humanity's exploration of space. By continuing to invest in the development and operation of Starship, SpaceX is helping to pave the way for a new era of space exploration and discovery.

The environmental impact of Starship technology

While the potential benefits of Starship technology are numerous, it is also important to consider its environmental impact. Spacecraft like Starship have the potential to cause environmental damage in a number of ways, including pollution and the destruction of habitats.

One potential concern is the pollution caused by the launch and operation of Starship. Rockets release large amounts of pollutants into the atmosphere, including carbon dioxide, nitrogen oxides, and other greenhouse gases. These emissions contribute to climate change and air pollution, and can have negative impacts on both human health and the environment.

Starship minimizes the effect on Earth's atmosphere by its use of methane and oxygen ("methalox") for propulsion. Although methane is a powerful (but short-lived) greenhouse gas, when burned the waste products are only CO_2 and water vapor, along with a small amount of nitrogen oxides. Reentry plasma and the heat from rocket exhaust also produce some nitrogen oxide, but not a large amount. Nitrogen oxide, when placed high in the atmosphere, could harm the ozone layer. If we take all this into account, Starship's use of methalox as a propellant should not cause serious problems with our atmosphere.

Another potential concern is the impact of Starship on the habitats and ecosystems of other planets. As humans begin to colonize other planets, there is a risk that we could unintentionally introduce invasive species or disrupt the natural balance of ecosystems. Additionally, the process of terraforming planets, or making them habitable for humans, could require significant alteration of the planet's natural environment, potentially causing irreparable damage.

However, there are also potential ways in which Starship technology could benefit the environment. For example, space-based solar power could provide a clean, renewable source of energy for Earth. Additionally, the development of space-based mining and manufacturing could help reduce the environmental impact of resource extraction and production on Earth. Space-based industries may create pollution, but should not affect us here on Earth.

To mitigate the potential environmental impacts of Starship technology, it is important to consider these concerns during the design and operation of spacecraft. This could include the use of cleaner fuels or the development of new propulsion systems that produce fewer pollutants. Additionally, it may be necessary to conduct careful assessments of the potential environmental impacts of colonization and terraforming activities.

The environmental impact of Starship technology is a complex issue that requires careful consideration. While the potential benefits of the technology are significant, it is important to weigh these benefits against the potential environmental impacts and work to mitigate any negative effects. By taking a thoughtful and responsible approach to the development and operation of Starship technology, we can help ensure that our exploration of space does not come at the expense of the environment.

The potential risks and dangers of Starship travel

While Starship technology has the potential to revolutionize space travel, it is not without risks and dangers. The challenges of long-duration spaceflight and the harsh conditions of space pose numerous risks to human health and safety.

One of the biggest risks associated with Starship travel is exposure to radiation. The high-energy particles found in space can damage DNA and increase the risk of cancer and other diseases. Additionally, long periods of weightlessness can cause muscle and bone loss, cardiovascular problems, and other health issues.

Another potential risk is the possibility of equipment failure or malfunction. The complex systems required to keep a Starship functioning properly can be susceptible to errors, and a single failure could have catastrophic consequences. Additionally, the risks of collisions with space debris or other objects in space could pose significant dangers to the crew and the spacecraft.

Other risks associated with Starship travel include psychological stress and isolation, as well as the potential for accidents during landing or takeoff. These risks must be carefully considered and mitigated through rigorous training and safety protocols.

Despite these risks, the potential benefits of Starship technology may outweigh the dangers. With proper planning and preparation, it may be possible to mitigate many of the risks associated with long-duration spaceflight. Additionally, the knowledge gained from Starship missions could help us better understand the universe and our place in it, leading to new scientific discoveries and advancements.

Overall, the potential risks and dangers of Starship travel are significant, but they should not be seen as insurmountable obstacles. Through careful planning, research, and training, we can help ensure the safety and success of Starship missions, and pave the way for a new era of space exploration and discovery.

"I say something, and then it usually happens. Maybe not on schedule, but it usually happens."

Elon Musk

Chapter 6: The Future of Starship

The potential for Starship to transform space exploration and human civilization

The development of Starship technology has the potential to transform space exploration and human civilization in numerous ways. With the ability to travel farther and faster than ever before, Starship could help us unlock new frontiers in space and expand our understanding of the universe.

One of the most exciting possibilities of Starship technology is the potential for human colonization of other planets. With the ability to transport large numbers of people and supplies, Starship could make it possible to establish permanent settlements on Mars, the Moon, and other celestial bodies. This could lead to the development of new industries and economies, as well as new scientific discoveries and advancements.

Starship technology could also transform the way we approach space exploration. With its speed and efficiency, Starship could make it possible to send missions to the outer reaches of our solar system and beyond. This could lead to new discoveries about the origins of the universe and the possibility of extraterrestrial life.

Another potential benefit of Starship technology is the ability to address global challenges here on Earth. With its capacity for high-speed transport of people and cargo, Starship could make it possible to quickly respond to natural disasters, provide emergency medical supplies and equipment, and support scientific research in remote locations.

The potential for Starship to transform human civilization is not limited to space exploration and disaster response. The development of Starship technology could also lead to new advancements in transportation, energy, and other industries. For example, the materials and technologies used in building Starship could be applied to other fields, leading to new innovations and breakthroughs.

The potential for Starship technology to transform space exploration and human civilization is immense. As we continue to develop and refine this technology, we may be able to unlock new frontiers and address some of the most pressing challenges facing our world today. While there are certainly risks and challenges associated with Starship travel, the potential benefits make it an exciting and worthwhile endeavor.

The role of Starship in future space missions and colonization efforts

As the development of Starship technology continues, its potential role in future space missions and colonization efforts is becoming increasingly clear. With its ability to transport large numbers of people and cargo, Starship could be instrumental in establishing permanent human settlements on other planets and celestial bodies.

One potential use for Starship is to transport supplies and equipment for future space missions. This could include setting up bases on the Moon or Mars, where astronauts could live and work for extended periods of time. Starship could also be used to ferry resources and supplies to support these missions, such as food, water, and building materials.

In addition to supporting space missions, Starship could also play a critical role in human colonization efforts. With its capacity for transporting large numbers of people and cargo, Starship could make it possible to establish permanent settlements on other planets, such as Mars. This would require the development of self-sustaining colonies, where humans could live and work for extended periods of time, producing their own food, water, and other resources.

The potential for Starship to facilitate human colonization is particularly important, given the challenges facing our planet today. Climate change, overpopulation, and resource depletion are just a few of the issues that could be addressed through the establishment of self-sustaining colonies on other planets. These colonies could provide a new home for humanity, as well as a way to preserve our species in the event of a catastrophic event on Earth.

Starship can also be used to develop a space infrastructure, including asteroid mining, space factories, and even entire habitats. Our space infrastructure is very limited, as we can only loft smaller satellites weighting hundreds of pounds. How much different will our orbital world become when we can put thousands of pounds to work for us? Starship gives us a capability we have never had before, to launch and develop truly large structures in orbit of our planet and beyond.

However, the role of Starship in future space missions and colonization efforts is not without challenges. The technology required to establish self-sustaining colonies is complex, and there are many risks associated with long-term space travel and settlement. Additionally, there are questions about the ethics and responsibility of colonizing other planets and celestial bodies, particularly if this comes at the expense of indigenous life forms or ecosystems.

Despite these challenges, the potential for Starship to facilitate future space missions and colonization efforts is immense. As the technology continues to develop, we may be able to unlock new frontiers in space and establish new homes for humanity on other planets. With careful planning and responsible stewardship, Starship could be a powerful tool for expanding our understanding of the universe and preserving our species for generations to come.

The challenges and opportunities of a Starship-enabled future

As Starship technology continues to develop, it presents both challenges and opportunities for the future of space exploration and human civilization. While Starship has the potential to revolutionize space travel and enable human colonization of other planets, there are also significant obstacles to overcome in order to realize this vision.

One of the key challenges of a Starship-enabled future is the cost and complexity of building and operating these spacecraft. While SpaceX has made significant progress in reducing the cost of space travel through the development of reusable rockets and spacecraft, Starship is still a complex and expensive technology to build and maintain. In order to make Starship a viable option for space travel and colonization, significant investments will need to be made in research, development, and infrastructure.

Another challenge of a Starship-enabled future is the potential for environmental impact. As we explore and colonize other planets, we must be mindful of the impact that our activities may have on the local ecosystems and environments. This requires careful planning and responsible stewardship to ensure that our exploration and colonization efforts do not have unintended consequences for the planets and celestial bodies we visit.

Despite these challenges, a Starship-enabled future also presents many opportunities. One of the most significant of these is the potential for human exploration and colonization of other planets. This could open up new opportunities for scientific research, as well as provide a new home for humanity in the event of a catastrophic event on Earth.

Another opportunity presented by a Starship-enabled future is the potential for space tourism. As the cost of space travel decreases, it may become possible for more people to experience the thrill of spaceflight. This could open up new opportunities for entertainment, education, and adventure.

Finally, a Starship-enabled future could also have significant economic benefits. As we explore and colonize other planets, we may discover new resources and opportunities for commerce. This could create new jobs and industries, as well as drive innovation and economic growth.

A Starship-enabled future presents both challenges and opportunities for the future of space exploration and human civilization. While there are significant obstacles to overcome, the potential benefits of this technology are immense. With careful planning and responsible stewardship, we may be able to unlock new frontiers in space and establish a bright future for humanity in the cosmos.

"There's a silly notion that failure's not an option at NASA. Failure is an option here. If things are not failing, you are not innovating enough."

Elon Musk

Chapter 7: Conclusion

The promise and potential of Starship technology

The promise and potential of Starship technology is something that has captivated the imaginations of scientists, engineers, and enthusiasts for many years. The idea of traveling through space on a massive, self-sustaining spacecraft is a dream that has inspired countless works of science fiction, but it is also an idea that is quickly becoming a reality.

The Starship, as envisioned by SpaceX, is a spacecraft designed to transport humans and cargo to destinations beyond Earth's orbit, including the Moon, Mars, and potentially even further. It is a fully reusable spacecraft, which means that it is capable of making multiple trips without the need for extensive repairs or refurbishment between flights.

One of the key features of the Starship is its massive size. With a length of 50 meters and a diameter of 9 meters, it is significantly larger than any spacecraft currently in operation. This size allows it to carry a large amount of cargo, including scientific equipment, supplies, and even habitats for human habitation.

Another important feature of the Starship is its reusability. By being able to make multiple trips without extensive repairs or refurbishment, it drastically reduces the cost of spaceflight. This is a key factor in making space exploration and colonization more accessible to a wider range of people and organizations.

The Starship is also designed to be self-sustaining, which means that it is capable of producing its own resources, such as water and oxygen, through processes like electrolysis. This is an important feature for long-duration space missions, as it reduces the need for resupply missions and allows for more extended missions.

The potential of the Starship technology goes beyond just space exploration and colonization. It also has the potential to revolutionize long-distance travel on Earth. For example, the Starship could be used to transport people and cargo around the world in a matter of hours, drastically reducing travel times and making global trade more accessible.

The promise and potential of Starship technology are immense. It has the potential to revolutionize space exploration, colonization, and even long-distance travel on Earth. While there are still many challenges to overcome before it becomes a reality, the Starship represents a significant step forward in humanity's quest to explore and inhabit the cosmos.

The need for continued innovation and development in space exploration

The exploration of space has always been a complex and challenging endeavor, requiring a significant investment of resources, time, and effort. From the first human spaceflight in 1961 to the recent launch of the Mars rover Perseverance and the James Webb Space Telescope, space exploration has continued to evolve and advance. Each new mission pushes the boundaries of human knowledge and capability.

However, as we look to the future, it is clear that continued innovation and development in space exploration will be essential if we are to continue making progress in our understanding of the universe and our ability to explore it. There are several key reasons why this is the case.

Firstly, space exploration has the potential to yield enormous scientific and technological benefits. For example, space research has led to significant advances in areas such as materials science, energy production, and medical technology. Continued innovation in space exploration will allow us to build on these achievements and unlock new discoveries and technological breakthroughs.

Secondly, space exploration is critical for our understanding of the universe and our place in it. As we explore new planets, moons, and asteroids, we are learning more about the formation and evolution of the solar system and the universe as a whole. This knowledge can help us answer some of the most fundamental questions about our existence, such as how life on Earth began and whether we are alone in the universe.

Thirdly, space exploration has the potential to help us address some of the most pressing challenges facing humanity. For example, by developing new technologies for space travel and habitation, we can also develop new ways to address issues such as climate change, resource depletion, and overpopulation. In addition, space exploration can help us prepare for potential global catastrophes, such as asteroid impacts or solar storms.

Despite the enormous potential of space exploration, there are also significant challenges that must be overcome. These challenges include the high cost of space exploration, the difficulty of developing technologies that can withstand the harsh conditions of space, and the risk to human life that comes with space travel.

To address these challenges, continued investment in space exploration is essential. This investment should focus on developing new technologies that can make space exploration more cost-effective and safe, as well as expanding our scientific knowledge and understanding of the universe. In addition, we must also work to build international partnerships and collaborations to ensure that space exploration is a global effort that benefits all of humanity. Every country has a part to play, and a responsibility to their citizens to help build our future in space.

The need for continued innovation and development in space exploration is clear. By investing in this critical area of human endeavor, we can unlock new scientific and technological breakthroughs, deepen our understanding of the universe, and address some of the most pressing challenges facing humanity.

The importance of the new Space Age in shaping the future of humanity

The new Space Age, characterized by the rapid development of commercial spaceflight and the increasing interest in space exploration and colonization, represents a critical juncture in the history of humanity. This new era has the potential to shape our future in profound ways, from advancing our scientific knowledge to transforming our economy and society.

One of the most important ways in which the new Space Age is shaping the future of humanity is through the advancement of scientific knowledge. By exploring new planets, moons, and asteroids, we are gaining a deeper understanding of the universe and our place within it. This knowledge can help us answer fundamental questions about our existence, such as how life on Earth began and whether we are alone in the universe. In addition, the new Space Age is also leading to the development of new technologies that have the potential to transform many aspects of our lives, from energy production to medical treatments.

Another important way in which the new Space Age is shaping the future of humanity is through the transformation of our economy and society. As commercial spaceflight becomes more commonplace, it has the potential to create entirely new industries and markets, from space tourism to asteroid mining. These industries can drive economic growth and create new jobs, while also providing new opportunities for innovation and entrepreneurship. In addition, the development of new technologies for space exploration and habitation can also have a significant impact on our society, by providing solutions to challenges such as climate change, resource depletion, and overpopulation.

The new Space Age also has the potential to foster international cooperation and collaboration. Space exploration has always been a global endeavor, with many countries and organizations working together to achieve common goals. The increasing involvement of commercial entities in space exploration has the potential to further drive this collaboration, by bringing together stakeholders from across the world to work towards common goals. This cooperation can help us address global challenges such as climate change and geopolitical tensions, while also promoting peace and understanding among nations. Every nation has a stake in the Space Age, whether they act on it or not.

However, there are also challenges and risks associated with the new Space Age. These include the potential for space debris to cause damage to spacecraft and infrastructure, the risk of collisions between spacecraft, and the potential for the militarization of space. It is essential that we work to address these challenges and mitigate these risks, while also ensuring that the benefits of the new Space Age are accessible to all of humanity.

The new Space Age represents a critical juncture in the history of humanity, with the potential to shape our future in profound ways. By advancing our scientific knowledge, transforming our economy and society, fostering international cooperation and collaboration, and addressing the challenges and risks associated with space exploration, we can ensure that the new Space Age is a force for good that benefits all of humanity.

Addendum

https://www.spacex.com/
SpaceX
SpaceX designs, manufactures and launches the world's most advanced rockets and spacecraft. The company was founded in 2002 by Elon Musk to revolutionize space transportation, with the ultimate goal of making life multiplanetary. SpaceX has gained worldwide attention for a series of historic milestones.

https://www.spacex.com/media/making_life_multiplanetary_transcript_2017.pdf
MAKING LIFE MULTIPLANETARY - SpaceX
This document is an abridged transcript of Elon Musk's presentation at the 68th International. Astronautical Congress on September 28th, 2017 in Adelaide, Austraila

https://www.spacex.com/media/starship_users_guide_v1.pdf
Starship Users Guide – SpaceX 6 pages
SpaceX's Starship system represents a fully reusable transportation system designed to service Earth orbit needs as well as missions to the Moon and Mars.

https://www.spacex.com/vehicles/starship/
Starship – SpaceX
STARSHIP - SERVICE TO EARTH ORBIT, MOON, MARS AND BEYOND

https://www.spacex.com/updates/
Updates - SpaceX
Dec 8, 2022 — NASA announced it has modified its contract with SpaceX to further develop the Starship human landing system. Initially selected to develop a lunar lander capable of carrying astronauts between lunar orbit and the surface of the Moon as part of NASA's Artemis III mission—marking humanity's first return to the Moon since the Apollo program's final mission in 1972—SpaceX will now support a second human landing demonstration as part of NASA's Artemis IV mission.

https://en.wikipedia.org/wiki/SpaceX_Starship
SpaceX Starship - Wikipedia
The SpaceX Starship is a fully reusable super heavy-lift launch vehicle under development by SpaceX. Standing at 120 m (390 ft) tall, it is designed to be the tallest and most powerful launch vehicle ever built, and the first capable of total reusability.

https://www.popularmechanics.com/science/a32052844/spacex-starship-user-guide-payload/
SpaceX's Starship Can Lift a Lot More Than We Thought
Popular Mechanics
Apr 6, 2020 — It's secretly swole... What's inside Elon Musk's quickstart guide to space travel? Well, mostly it's technical data, and even that's more like background information and high-level overviews than how to "use" anything. The concrete takeaways are intended for just one group: "Potential Starship customers can use this guide as a resource for preliminary payload accommodations information."

https://www.nytimes.com/article/elon-musk-starship.html
What Is Starship? SpaceX Builds Its Next-Generation Rocket
The New York Times
Feb 11, 2022 — NASA wants to use it to land American astronauts on the moon. The Pentagon wants to use it to whisk military cargo around the world in minutes. Astronomers, satellite companies and aspiring space tourists are eyeing its potential to drastically slash the cost of getting to space.

https://arstechnica.com/science/2022/02/heres-what-im-hoping-to-learn-from-this-weeks-starship-presentation/
Ars Technica
Feb 8, 2022 — The Starship will be "Ship 20." There haven't been 19 previous Starship prototypes, but there have been a lot. And this ship will be stacked on "Booster 4." It will make for an impressive backdrop, but will either of these vehicles take flight?

https://science.nasa.gov/science-news/science-at-nasa/2000/ast13nov_1
Breathing Easy on the Space Station - NASA Science
Nov 12, 2000 — The oxygen that people breathe on Earth also comes from the splitting of water, but it's not a mechanical process. Plants, algae, cyanobacteria and phytoplankton all split water molecules as part of photosynthesis -- the process that converts sunlight, carbon dioxide and water into sugars for food. The hydrogen is used for making sugars, and the oxygen is released into the atmosphere.

https://everydayastronaut.com/
Everyday Astronaut - Bringing space down to Earth for everyday people! Find out everything you need to know about upcoming and previous rocket launches.

https://www.youtube.com/watch?v=LbH1ZDImal8
Is SpaceX's Raptor engine the king of rocket engines?
YouTube · Everyday Astronaut
May 25, 2019

https://dearmoon.earth/
Meet the dearMoon Crew!
The first civilian mission to the Moon is planned to take place in 2023. The rocket developed by Elon Musk's SpaceX will make a week-long journey to the Moon and back. In 2018, Japanese entrepreneur Yusaku Maezawa purchased all the seats aboard this rocket. Wishing to give as many talented individuals as possible the opportunity to go, he announced in March 2021 his plans to choose 8 crew members from across the world. Here are the schedule and flight plans for the first civilian lunar orbital mission, the dearMoon project.

https://polarisprogram.com/
Polaris Program
The program consists of up to three human spaceflight missions that will demonstrate new technologies, conduct extensive research and ultimately culminate in the first flight of SpaceX's Starship with humans on board.